WORM COMPOSTING HANDBOOK

**A Comprehensive Approach To
Successful Vermicomposting:
Creating Sustainable Gardens
Through Vermiculture**

GATLIN ARES

Table of Contents

INTRODUCTORY

The natural process of composting transforms organic waste (such as food scraps and yard trimmings) into a soil conditioner rich in nutrients. Bacteria, fungus, and other microbes involved in decomposition aid in this process by converting organic matter into compost, a humus-like material.

Composting is based on the premise that these microbes can thrive in a controlled setting. This is a typical use for compost bins or piles. Composting involves the gradual breakdown of organic materials such as food scraps (e.g., peels, seeds, and rinds), yard debris (e.g., dead grass and leaves), and other similar items.

The ideal ratio of nitrogen-rich green materials to carbon-rich brown materials, along with moisture and air, are necessary for composting. Brown materials offer energy in the form of carbon, whereas green materials supply nitrogen—an element crucial to microbial growth. To speed up the decomposition process, it's a good idea to turn or stir the compost pile frequently to introduce air.

The formation of a nutrient-rich soil amendment that enhances soil structure and fertility, the reduction of kitchen and yard waste going to landfills, and the promotion of environmental sustainability through recycling organic materials are all advantages of composting. Both small-scale farmers and those who tend gardens at home can benefit from this method.

CHAPTER ONE
Worm Composting And Its Advantages

One kind of composting that makes use of worms—usually red wigglers (Eisenia fetida)—to decompose organic materials is worm composting, sometimes called vermicomposting. This approach has multiple advantages:

• Worms are fantastic decomposers because of how efficiently they work. Their nutrient-rich castings (worm feces) are a great natural fertilizer, and they make their living eating organic garbage.

• Compost of Superior Quality: Worm castings, often called vermicast, are a treasure trove of beneficial microbes, nitrogen, and phosphorus. This process produces compost of superior quality, which has the potential to greatly enhance soil fertility and structure.

• Worm composting is an efficient method of recycling organic waste, including food scraps, with the goal of reducing waste. It contributes to trash reduction initiatives by keeping organic materials out of landfills, which in turn lowers the total amount of waste.

• Worm composting is ideal for people who live in cities or who have a small yard because it doesn't require a lot of room. You can compost all year round with worm bins, whether you keep them indoors or on a balcony.

• Worm composting systems typically require little in the way of upkeep.

Worms, once established, work nonstop to decompose organic materials. Keeping an eye on the humidity and feeding it every so often should be enough.

• Worm bins do not emit unpleasant scents when kept in good condition. In contrast to the disagreeable odors that can arise during anaerobic decomposition, vermicomposting is an aerobic process that takes place in an oxygen-rich environment.

• Beneficial for Education: Worm composting is a fun and instructive hobby that adults and kids alike may enjoy. It elucidates the link between waste minimization and ecological sustainability as well as the natural process of decomposition.

- Complementing conventional composting processes, worm composting offers a versatile alternative. For lesser quantities of organic waste or food scraps, worm composting is a great alternative to traditional composting, which is better suited to bigger yard waste.

Worm composting is an eco-friendly and long-term solution to the problem of organic waste management, with the added benefit of creating high-quality compost for use in landscaping and plant care.

Comprehending Worms' Function In The Composting Process

Crucial to the process of composting, worms hasten the breakdown of organic materials and aid in the production of nutrient-rich compost. Red wiggler worms (Eisenia fetida) and other worms serve the following crucial roles in composting:

• Ingesting Organic Debris: Worms have an insatiable appetite for organic garbage. They feed on a wide range of organic waste products, such as the rinds and seeds of fruits and vegetables, as well as coffee grounds, eggshells, and other food scraps. As they consume these elements, they reduce them to tiny pieces.

• Worms' gizzards, which are muscular organs, mechanically break down food particles by grinding and pulverizing them. Mechanical breakdown occurs as organic material passes through the gizzard, which speeds up the decomposition process even further.

• The microbial activity of worms is a key component in their decomposition of organic materials. Enzymes produced by these microbes simplify complicated organic substances, which aids in the decomposition process.

• Worm digestion produces nutrient-rich castings, also called vermicast or worm castings, which enrich the soil and food supply. Essential nutrients such as micronutrients, nitrogen, phosphorus, and potassium can be found in abundance in these castings. Worms deposit nutrient-rich pathways as they tunnel into compost.

• Worms improve aeration by making channels as they move through compost, which allows air to circulate more freely. Because it supplies oxygen to the microbes that break down organic matter, this is vital to the composting process. Anaerobic decomposition produces unpleasant odors; however, well-aerated compost promotes aerobic conditions and keeps this from happening.

• Worms play a key role in maintaining a consistent pH level in the composting process. Their feces has a pHneutralizing effect on compost, which helps keep the breakdown bacteria in a stable environment.

• A More Diverse Community of Microbes: Worms in compost promote microbial diversity. The variety of microbes in composting aids in decomposing various organic components, leading to a compost that is both more stable and rich in nutrients.

• The byproduct of vermicomposting, worm castings, are terrific for amending soil. Garden soil that has it added to it has better structure, holds more water, and plants are able to access more nutrients.

Worms improve composting by eating, decomposing, and turning organic waste into castings that are rich in nutrients. Vermicomposting is an excellent and long-term waste recycling strategy since their actions improve the compost's overall efficiency and quality.

CHAPTER TWO
Making An Informed Worm Selection

To get the most out of your vermicomposting efforts, it's crucial to choose the correct worms. Red wigglers (Eisenia fetida) are the most often suggested worm species for composting, however not all worms are good for it. Before you buy worms to use in vermicomposting, think about the following:

• The vermicomposting method most commonly employs red wigglers, scientifically known as Eisenia fetida. Worms devour organic waste at an alarming rate, multiply rapidly, and thrive in the cramped conditions of a worm bin. Composting food scraps is a perfect fit for them because of their capacity to flourish in organic materials that is degrading.

• Keep Earthworms Out of Your Garden: Even while earthworms are essential for healthy soil, they aren't the ideal option for vermicomposting containers. Worms from garden soil, or earthworms, are naturally burrowers and could not do well in a worm bin due to the cramped quarters and other environmental constraints. Composting is a better habitat for red wigglers because they are surface dwellers.

• Vermicomposting also makes use of European nightcrawlers, scientifically known as Eisenia hortensis. They may be more appropriate for larger-scale composting systems due to their size compared to red wigglers. But because of their versatility and rapid reproduction rates, red wigglers are frequently chosen for home vermicomposting on a smaller scale.

• Worm Source: To acquire healthy, disease-free worms, buy them from a trustworthy vendor. Excellent resources can be found at nearby gardening stores, on the internet, or from other vermicomposters. You should not bring worms you find in the wild into a worm bin because they might not be healthy enough or because they might not adapt well to the regulated conditions there.

• Population Size: Think about how much organic waste you produce and the size of your vermicomposting bin. Even if you start with a small population of red wigglers, they will multiply fast enough to fill your composting system.

• Verify that the worms you select can withstand the temperature range in which your vermicomposting system operates. For instance, red wigglers do best between 13 and 25 degrees Celsius (55 and 77 degrees Fahrenheit).

Vermicomposting is all about worms, but it's also about keeping the composting system in the correct circumstances, such as with the right bedding, enough moisture, and a balanced diet of organic waste. Composting your food wastes into a nutrient-rich soil amendment is possible with the correct worms and the correct conditions.

Choosing An Appropriate Container

If you want your vermicompost to be as efficient as possible, you need to make sure the worms are in an ideal environment, which includes picking the right container. Some things to think about when choosing a container are:

1. Size:

• Think about the space you have and how much organic waste you produce. For households with moderate kitchen scraps, a small container might be enough; for bigger quantities of waste, larger containers or even multiple bins would be necessary.

2. Material:

• Vermicomposting often makes use of plastic bins. In addition to being easy to work with, they are lightweight and robust. Use only strong plastic that does not include any dangerous chemicals.

• You also have the option of using wooden bins. Even though they're insulating, they could be bulkier and more likely to wear out quickly.

Metals can corrode and hurt the worms, so don't use them.

3. Lid:

• Worms love a dark habitat, which a lid may offer, and it also helps maintain moisture levels. Odor control is another benefit. Pick a receptacle that fits snugly on top.

4. Air circulation:

• Proper ventilation is essential for keeping the compost in an aerobic environment. If the container you're considering doesn't already include holes for air circulation, you should try to find one that does.

5. Systems with Multiple Trays or Stackability:

• The completed compost can be more easily harvested using stackable or multi-tray systems. As the worms forage for new food sources higher up, they leave the completed compost in the bottom trays. Worms can be more easily separated from compost with this design.

6. Reducing Humidity:

• A method to regulate moisture is an important feature of an ideal vermicomposting container. Some examples of this would be a tray to catch rainwater or holes in the bottom for drainage.

7. Shade of Black:

• Light-colored containers lose less heat than dark-colored ones. A dark container might help keep the vermicomposting worms in the ideal temperature range, as they are temperature sensitive.

8. Collecting Made Simple:

• Think about the ease of harvesting the final compost. Sliding trays or other easy-access places are available in some systems.

9. How long does it last?

• Select a sturdy container that can endure the composting process. It is important to use containers that can resist dampness and occasional stirring for vermicomposting.

10. Money needed:

• You may find vermicomposting containers in several pricing ranges. Before you buy a container, figure out how much money you have to spend.

Bins with covers made of plastic, worm bins sold at stores, or homemade wooden bins are some of the most common options for vermicomposting containers. The most important thing is to make sure the worms are comfortable and that you can easily manipulate their surroundings so that the composting process goes smoothly.

Making Things Perfect For Worms

When designing a vermicomposting system, it's important to keep in mind the worms' specific requirements for survival, reproduction, and composting efficiency. Some important things to keep in mind are:

1. Temperature:

• For the most popular worms used in vermicomposting, red wigglers, keep the temperature between 55 and 77 degrees Fahrenheit, or 13 and 25 degrees Celsius. The worms are delicate and will die if exposed to too much heat or too much cold.

2. Dew point:

• Use a wrung-out sponge to keep the bedding in the vermicomposting container continuously damp. The worms can't breathe or digest without enough moisture. Add water if the bin gets too dry; dry bedding materials, such as shredded newspaper or cardboard, can be added if it gets too damp.

3. Material for the Bed:

• Give the worms something to live on and a way to keep the moisture in by adding a layer of bedding. Shredded newspaper, cardboard, aged compost, and coconut coir are common materials for bedding. To make the worms feel at home, the bedding should be soft and fluffy.

4. Fresh air:

• Turn or fluff the bedding on a regular basis to make sure it gets enough air. This keeps the soil from being too compacted and gives the microbes and worms that aid in breakdown oxygen. The aeration process is further enhanced by the container's ventilation openings.

5. Keeping the pH Levels in Check:

• Keep the vermicomposting system's pH between 6.0 and 7.0, which is neutral to slightly acidic. Worms thrive in slightly acidic conditions. If the pH gets too high, add some agricultural lime or broken eggshells; if it gets too low, add some acidic materials, such coffee grounds, in little quantities.

6. Absence of light:

• Dark places are more appealing to worms since they are photosensitive. If you must store the bin outdoors, make sure it is in a shady spot and use a lidded container to create a gloomy environment.

7. Correct Nutrition:

• Provide the worms with a varied diet of food scraps from around the kitchen, but keep them away from things that are acidic, spicy, meat, dairy, or oily. To hasten the decomposition process, mince or grind food leftovers into smaller bits. To avoid stinking up the house and attracting fruit flies, bury the leftovers in the bedding.

8. Refrain from Overfeeding:

• No need to feed the worms too much. Red wigglers can subsist on leftovers equal to about half their weight per day. A decrease in worm health, anaerobic conditions, and unpleasant odors can result from overfeeding.

9. Ongoing Oversight:

• Check on the worms' moisture, temperature, and general well-being on a frequent basis. Make any necessary adjustments to the conditions to guarantee the worms' well-being and the vermicomposting process's success.

10. Collecting Composted Debris:

• To keep worm populations from getting out of hand, harvest completed compost on a regular basis to make room for them. Worms in a multi-tray system will leave the finished compost in the bottom tray and move on to the fresher bedding in the upper tray.

By keeping these things in mind and adjusting as necessary, you can make sure your worms are happy and healthy while getting the most out of your vermicomposting efforts.

CHAPTER THREE
Constructing Your Worm Trap

When preparing a worm bin for vermicomposting, it is important to provide the worms with an appropriate habitat, as well as the correct bedding, food, and environmental conditions. To get your worm bin up and running, follow these simple steps:

What You'll Need:

1. Container:

• To begin vermicomposting, select an appropriate container. Most people use a plastic container with a cover, but you may also buy worm bins or build your own wooden container.

2. Material for the Bed:

• Cardboard, shredded newspaper, coconut coir, or a combination of the three. The bedding should be damp, but not soaked, before use.

3. Eisenia fetida, most commonly known as red wigglers:

• Get your hands on some red wiggler worms from a trustworthy vendor. Your bin's capacity and the quantity of organic waste you produce will determine the optimal worm population.

4. Air circulation:

• To make sure there's enough airflow in a container that doesn't already have it, you can drill tiny holes in the top or sides.

5. Reducing Humidity:

• To keep the soil at the ideal moisture level, you can use a spray bottle to apply water as needed.

6. pH Modifiers:

• If needed, you can modify the pH with crushed eggshells or agricultural lime.

Methods for Establishing a Worm Bin:

1. Find a suitable container:

• To allow air to circulate, puncture the plastic bin's lid and sides with tiny holes. If the bin gets too wet, set it on a tray or put it in a shallow pan.

2. Make the Bed:

• Make little pieces of cardboard or newspaper. Bring the bedding material to a wet consistency, like a wrung-out sponge.

3. Pile on the Bedding:

• Pour enough wet bedding material to cover the base of the container by four to six inches.

4. Let Worms In:

• Lay the bedding down and then add the red wiggler worms. If you disperse them equally, they'll start to dig themselves a home in the bedding.

5. Use Bedding to Cover:

• Sprinkle more damp bedding over the worms. In addition to creating a cozy atmosphere, this aids in keeping things dark.

6. First Meals:

• To start, put a little bit of food waste in the bin. To keep the bedding odor-free and worm-friendly, bury the waste.

7. Track and Modify:

• Always make sure the worms are dry, warm, and healthy by checking on them at regular intervals. If the bedding is too wet, add water; if the water is too dry, add more bedding.

8. Refrain from Overfeeding:

• Begin with a little quantity of food and progressively raise it as the worm population expands. To keep problems like stinky fish and fruit fly infestations at bay, don't overfeed your fish.

9. Collecting Worm Manure:

• Collect the composted material as it settles. Compost is simply removed from multi-tray systems when worms move to new bedding.

10. Take Care of the System:

• Add air circulation by turning or fluffing the bedding on a regular basis. From time to time, check the pH and make any required adjustments with agricultural lime or broken eggshells.

Successful vermicomposting systems convert food scraps into nutrient-rich compost for plants with little more than following these instructions and keeping an eye on the conditions in your worm bin.

Maintaining A Worm Diet

If you want your vermicomposting system to work and your worms to stay healthy, you need to feed them a proper diet. A wide variety of organic materials are ideal for the red wiggler worm's (Eisenia fetida) diet. A few pointers on what to feed your worms:

Worm-Friendly Diet:

1. Veggie and Fruit Byproducts:

• These are some great options: banana peels, apple cores, orange rinds, carrot tops, and any other leftover fruit or vegetable scraps.

2. Used Coffee Filters and Grounds:

• You can enrich the worm bin with nitrogen by adding used coffee grinds and filters.

3. A delicate balancing act:

• Crushed eggshells, rinsed to eliminate egg residue, contribute calcium to the compost and help balance the pH.

4. Nutrient-Rich Fruit Skins:

• Worms can eat the peels of fruits that aren't citrus, like berries, grapes, and melons.

5. Peeled vegetables:

• The inclusion of vegetable peels, such as those from cucumbers, potatoes, and others, is highly appreciated.

6. Steep Leaf and Tea Bags:

• You can put used tea bags (after you remove the staples) and tea leaves in the worm bin because they are compostable.

7. Debris of Paper and Media:

• Worms can be fed a tiny amount of unbleached, shredded paper and cardboard, which also serves as bedding.

8. How Compost Ages:

• Since completed compost includes helpful microbes, it can be put in tiny amounts on occasion.

Avoid These Foods:

1. Pomegranate Juice:

• When consumed in excess, the acidity of citrus fruits and peels can kill worms.

2. Garlic and Onions:

• These can make the bin smell bad, which worms might not like.

3. Seasonal Spices:

• Worms are delicate and can't handle strong flavors, so cut out the hot stuff.

4. Items containing dairy and meat:

• Keep away from meat and dairy items since they bring in pests and make things smell bad.

5. Rich Cuisine:

• Worms may have trouble navigating greasy or oily foods because they generate a slippery surface.

6. Refined Meals:

• Stay away from processed foods that are heavy in salt, additives, or preservatives.

Feeding Suggestions:

1. Mix or Mince Food:

• To hasten the decomposition process and make the meal more digestible for worms, chop or mix food wastes into smaller bits.

2. Bury the Meat:

• To keep the worms in a healthy habitat and to keep fruit flies at bay, burrow the leftovers in the bedding.

3. Maintain a Moderate Diet:

• Begin feeding the worms a little bit at a time and add more as they multiply. Overfeeding can cause unpleasant smells and other problems, so try not to do it.

4. Keep an eye on the breakdown:

• Keep track of the worms' rate of meal digestion. Feed less frequently until they make up for lost time if there is an excess of leftovers.

5. Keep Brown and Green Materials in Balance:

•"Green" materials, such as food scraps, are high in nitrogen; "Brown" materials, such as bedding, are rich in carbon. This ratio should be maintained for optimal composting.

If you want your worms to be healthy and prolific, producing compost that is rich in nutrients for your plants, you need provide them with a varied and balanced diet while avoiding certain things.

CHAPTER FOUR
Worm Castings For Harvesting

Worm castings, or vermicast or worm compost, are an essential part of the vermicomposting process. You may feed your plants organic fertilizer that is rich in nutrients by using these castings. To get the most out of your worm bin, follow these steps:

When the Castings Are Properly Prepared for Harvest:

1. Drab, Crunchy Feel:

• The granular texture of castings, which typically range from dark brown to black, is reminiscent of rich, healthy soil.

2. A Natural Aroma:

• A nice, earthy aroma should emanate from the castings. Anaerobic conditions or overfeeding might be the cause of a foul odor in the substance.

3. Migration of Worms:

• Worms would often seek out newer bedding rather than linger on old casts. A possible sign that the castings are ready for collection is a cluster of worms in one spot.

Worm Castings Harvesting Procedures:

1. Quit Feeding:

• A few weeks prior to harvesting, refrain from adding any new produce to the compost bin. Worms will be more motivated to complete digesting of the present material if this is done.

2. Transfer Raw Materials:

• Put undigested food scraps and other unprocessed materials on one side of the trash can. You can put this stuff back in the trash to make it break down even more.

3. Form an Additional Area for Feeding:

• Place new food scraps and bedding on the side of the bin that is empty. This makes room for the worms to go to a different area.

4. Await the Migration of Worms:

• The worms will need a few weeks to move from the completed castings to their new feeding area.

5. Gather Encased Parts:

• After the worms have moved on, you can remove their finished castings by scooping them up from the side they were on. To gently collect the castings, you can use your hands or a little hoe.

6. Worm Separation:

• The collected material can be placed on a mound with a light source overhead to separate any leftover worms from the castings. The worms will dig themselves into the center of the mound because they are photosensitive. The majority of the worms will be in the upper layer, so scoop it out.

7. Get Worms Out of the Bin:

• Add some new bedding and leftover food to the bin's feeding area, then return the worms you just caught.

8. Do it again if needed:

• You can start the process over by making a new feeding zone and giving the worms another chance to move if you find a large number of them in the gathered castings.

9. Keep Castings for Future Use:

• Either immediately plant the collected castings in your garden or store them in a container with enough air circulation.

You can use castings as a top treatment, mix them with potting soil, or scatter them around plants.

Some pointers:

• Castings are ready for harvest when they are uniformly dark and crumbly; nevertheless, a few worms in the mix are perfectly OK and will not hurt your plants.

• A multi-tray arrangement might make harvesting a little less of a hassle. The final castings may be simply scraped out of the lower trays as the worms climb upwards to locate fresh bedding.

Following these methods will help you harvest worm castings in an effective way. This will allow you to return vital nutrients to your garden while also giving the worms a place to continue composting.

Common Problems And Their Solutions

Although vermicomposting is usually a simple operation, problems can nevertheless develop in any system. Some typical issues with worm bins and ways to fix them are as follows:

1. Reputable Smells:

• Cause: Anaerobic conditions, brought on by overfeeding or too much moisture, typically manifest as foul odors.

Approach:

• For the time being, stop increasing food.

- Rotate the mattress to make sure it gets enough air.

- Layer in some dry items for the bedding, such as cardboard or shredded newspaper.

- Verify the amount of moisture and make any required adjustments.

2. Pests Like Fruit Flies:

- Fruit flies and other pests might be attracted to ripe or exposed food.

Approach:

- Bury leftover food in the bedding to keep it out of sight.

- Make sure the top is tightly screwed on.

• To make your garden less appealing to pests, don't overfeed it.

3. The Worms' Effort to Flee:

• Root: Unwanted environmental factors within the container, including high temperatures, high humidity, or acidity.

Approach:

• Please make any necessary adjustments to the moisture levels after checking.

• Make sure there is enough oxygen.

• Verify the pH and make any required adjustments.

4. Gradual Degradation:

• Reason: If there isn't enough microbe activity, meals can not break down properly.

Approach:

• To make sure there's an adequate amount of carbon to nitrogen, add more "browns" (dry ingredients like crushed paper).

• Grind or chop leftover food into tiny bits.

• Verify and modify the moisture levels.

5. Is It Too Damp or Too Dry?

• Worm activity might be affected by inadequate moisture levels.

Approach:

• If it's too dry, add water; if it's too wet, add dry bedding materials.

• Get it to a consistency that's like a wrung-out sponge in terms of wetness.

6. Acid-Base Discord:

• Worm health might be affected by an environment that is too acidic or too alkaline.

Approach:

• To raise the pH, you can use agricultural lime or broken eggshells.

• To lower the pH, add a few drops of acidic ingredients, such as coffee grounds.

7. Too Few Worms:

• Reason: Decomposition could be slowed down if the worm population isn't sufficient to handle the waste.

Approach:

• Be careful not to feed the worms too much until their numbers increase.

• If necessary, think about adding additional worms.

8. Bedding with Compactions:

• The limitation of aeration occurs as a result of bedding compacting.

Approach:

• Regularly fluff or rotate the bedding to enhance air circulation.

- To avoid compacting, make sure to include additional dry bedding materials.

9. Mice of white coloration:

- Reason: In extremely damp environments, white mites may manifest.

Approach:

- Let the bedding air dry a bit.

- Be sure to regulate the moisture and aerate the area properly.

10. Bad Smell in Gathered Castings:

- Anaerobic conditions during decomposition could be the reason why harvested castings have a disagreeable odor.

Approach:

• Make sure that the composting process is adequately aerated.

• Before harvesting, let the casts dry.

In order to keep your vermicomposting system healthy and productive, it is important to regularly check on the worms and make any necessary adjustments to the bin conditions.

CHAPTER FIVE
The Use Of Worms In Combination With Conventional Composting

Vermicomposting, or worm composting, is an effective and all-encompassing approach of manage organic waste that combines worms with conventional composting techniques. Combining the two approaches allows you to take use of their respective strengths. Combining worm composting with more conventional methods is as easy as following these steps:

1. Arrange in Distinct Containers:

• Separate the worm composting bins or piles from the regular composting bins. You can administer each system separately in this way.

2. Garbage from Yards for Conventional Composting:

• For larger pieces of yard debris, like leaves, grass clippings, and plant trimmings, use conventional composting methods. These items work well in conventional composting bins or piles, however they could take a little longer to decompose.

3. Worm Bin for Food Scraps:

• Put food leftovers in the worm bin, including discarded fruit and vegetable skins, coffee grinds, and eggshells. These materials are easily broken down by worms.

4. Multiple-Tray Worm Bin System:

• One option is to use a worm bin system with multiple trays. You can transfer additional food scraps to another composting tray as the worms complete their work in one. Because of this, worm castings may be harvested and composting can continue indefinitely.

5. Combine the final goods:

• If you want to give your garden an extra boost of nutrients, mix regular compost with worm castings when they're both ready. Worm castings enrich the compost with microbes and increase nutrient availability, while conventional compost offers a wide variety of nutrients. By combining the two, you get the best of both worlds.

6. Bedding Made with Conventional Compost:

• Line your worm bin with finished conventional compost. Worms benefit from this because it increases the amount of organic material available for decay in their ecosystem.

7. Flip the Objects:

• Mix the worm bin with regular compost by turning the materials around. For instance, you can use the worm bin to start with food wastes, then move the finished materials to the regular compost, and vice versa.

8. Let Conventional Compost Air Out:

• If you want your traditional compost to break down aerobically, you need to turn it often and add air. This aids in avoiding anaerobic conditions and foul odors.

9. Worm Castings for Conventional Composting and Harvesting:

• Take out the worm castings from the worm bin on a regular basis and mix them with the ordinary compost. Because of this, the nutritional content and microbial diversity of conventional compost are both improved.

10. Make Smart Bin Placements:

• Do not forget to set up your worm bin and conventional composting pile in easily accessible areas. Put the worm bin somewhere shady or indoors, and put the conventional composting area next to your garden so you may easily reach the finished compost.

11. Opportunities for Education and Community Service:

• Whether you're in a community or school environment, teaching people how to compost using worms or the traditional way is a great way to get their hands dirty. Draw attention to the ways in which each approach differs and excels.

You can more effectively handle a wider variety of organic materials by combining worm composting with conventional composting. A more sustainable and well-balanced system for your garden is created when you combine methods that maximize nutrient cycling and reduce waste.

Worm Composting For Your Garden

Compost made from worm castings, an organic fertilizer rich in nutrients, has many uses in gardening. To get the most out of worm compost in your yard, try these methods:

1. Dressing the Top:

• Cover the soil in your garden with worm compost. Increased nutrient availability, water retention, and soil structure are all benefits of this top dressing.

2. Bring to a combined potting mixture:

• Combine worm castings with other ingredients such as coconut coir, perlite, and vermiculite to make a unique potting mix, or add worm compost to existing potting soil.

3. Mix for Starting Seeds:

• Add worm castings to a seed starting mix. The delicate, nutrient-rich setting fosters the growth of seedlings.

4. Performing transplants:

• When transferring trees or plants, be sure to include worm compost in the planting hole. This gives them a nutritional boost right from the start.

5. Enhancements to Soil:

• To improve the soil in your garden, mix worm compost with the soil you already have. Soil fertility, microbial life, and general soil health are all improved by this.

6. Leaf Tea: Compost

• Making compost tea is as simple as steeping worm castings in water. Foliar spray or soil drench plants with this nutrient-rich liquid.

7. The practice of mulching:

• Mulch your plants with a layer of worm compost. Soil temperature, moisture conservation, and nutrient release can all be aided by mulching.

8. Gardens for Plants and Flowers:

• Prior to planting, mix worm compost into flower and vegetable gardens. This ensures that the plant receives nutrients at all times during the growing season.

9. Gardening in Containers:

• When planting in containers, include worm compost in the soil mix. Beneficial for houseplants, the composition is light and rich in nutrients.

10. Enhancing the Pile of Compost:

• To speed up the decomposition process and increase microbial activity, you can add worm castings to regular compost piles.

11. Greenery: grass

• To promote healthy soil and increase grass growth, spread a thin layer of worm compost over lawns.

12. Household Plants:

• Add worm compost to soil for houseplants or use it as a decorative topping. Because of their gentle nature, worm castings are perfect for use in indoor gardening.

13. Shrubs and Perennials:

• Worm compost is a great addition to the soil around shrubs and perennials, especially when they are in bloom. Established plants love the slow-release nutrients.

14. Preventing Erosion:

• In places where erosion is a problem, mix in worm compost. With a better soil structure, soil erosion is less likely to occur.

15. Soil Amendment Without Acidity:

• Because of their generally neutral pH, a wide variety of plants can benefit from consuming worm castings. They won't have much of an effect on soil acidity, unlike other organic materials.

Some pointers:

• Apply worm compost in a quantity that corresponds to your plants' requirements. In general, it's safe to use, and the amount can be adjusted according to the type of plant and its growth stage.

• Perseverance: For continuous soil improvement, consistently incorporate worm compost.

To keep worm compost in good condition, store it in a cool, dry location.

To improve soil fertility and plant health in general, worm compost is a great, all-natural option. Consistent use can help with environmentally friendly and fruitful gardening methods.

Summary

Worm composting, also known as vermicomposting, is an eco-friendly and effective way to recycle organic waste and make nutrient-rich garden compost.

Red wiggler worms (Eisenia fetida) have innate abilities that can be used to transform organic waste (such as food scraps) into worm castings, a highly beneficial soil amendment.

For vermicomposting to be a success, you need to make sure the worms are in the correct environment, keep the composting material at the right moisture level, keep the carbon-to-nitrogen ratio balanced, and know how to fix common problems.

Decomposing organic materials, increasing microbial activity, and producing a nutrient-rich end product are all greatly assisted by worms.

To maximize the benefits of both systems and manage a wider range of organic materials, it is best to integrate worm composting with traditional composting methods.

Soil fertility, structure, and health are all enhanced when worm castings are used in various ways, such as a top dressing, added to potting mixes, or made into compost tea.

When it comes to sustainable gardening, vermicomposting is invaluable, and it's also good for the environment. Before you start vermicomposting, think about what your plants need, keep an eye on your worm bin to make sure it's healthy, and have fun making something useful out of your food scraps. Waste not, want not!

THE END